LET'S-READ-AND-FIND-OUT SCIENCE®

WHAT COLOR IS CAMOUFLAGE?

by Carolyn Otto

illustrated by Megan Lloyd

HarperCollinsPublishers

The illustrations in this book were done with watercolor and pen and ink
on Saunders Waterford Watercolor Paper.

The *Let's-Read-and-Find-Out Science* book series was originated by Dr. Franklyn M. Branley, Astronomer Emeritus and former Chairman of the American Museum–Hayden Planetarium, and was formerly co-edited by him and Dr. Roma Gans, Professor Emeritus of Childhood Education, Teachers College, Columbia University. Text and illustrations for each of the books in the series are checked for accuracy by an expert in the relevant field. For more information about the Let's-Read-and-Find-Out Science books, write to HarperCollins Children's Books, 10 East 53rd Street, New York, NY 10022.

Library of Congress Cataloging-in-Publication Data
Otto, Carolyn.
What color is camouflage? / by Carolyn Otto ; illustrated by Megan Lloyd.
p. cm. — (Let's-read-and-find-out science. Stage 1)
Summary: Explains how animal markings and colorings vary in response to their environment.
ISBN 0-06-027094-2. — ISBN 0-06-027099-3 (lib. bdg.)
ISBN 0-06-445160-7 (pbk.)
1. Camouflage (Biology)—Juvenile literature. [1.Camouflage (Biology). 2. Animal defenses.] I. Lloyd, Megan, ill. II. Title. III. Series.
QL767.077 1996 95-32173
591.57'2—dc20 CIP
AC

Typography by Al Cetta
1 2 3 4 5 6 7 8 9 10
❖
First Edition

WHAT COLOR IS

CAMOUFLAGE?

Mountain Lion

In the foothills near my house, a mountain lion might be hunting. He stalks across a dry hillside. His golden fur matches the grass.

A doe raises her head to sniff the air. Her fawn is quiet, and the lion passes by.

In my backyard, a robin could catch an insect that looks like a leaf.

Striped Skunks

After dark, a skunk and her babies might walk right down my street. My dog can see their bold white stripes, but he's learned to leave skunks alone.

Near my house, and all over the world, at every time of day or night, animals are hunting for food . . . or they are being hunted.

Animals must eat in order to live, and to live they sometimes have to hide. Animals hide in holes, burrows, and dens, in plants, underwater, and beneath rocks.

Great Horned Owl

Yellow-Bellied Marmots

Short-Horned Lizard

Some animals hide out in plain sight, but they are still very hard to see. The doe's coat and the mountain lion's fur blend into the colors surrounding them.

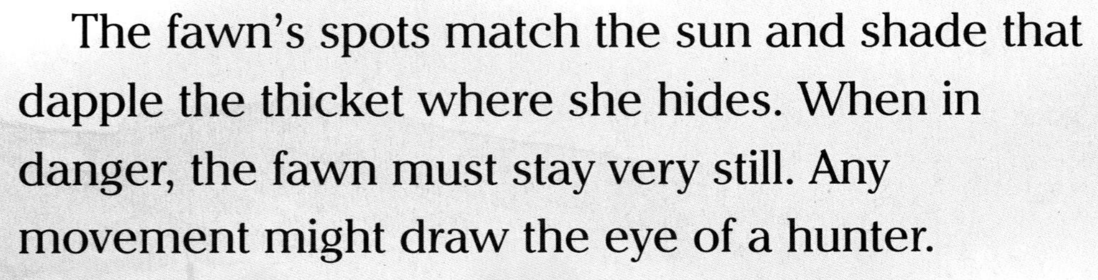

The fawn's spots match the sun and shade that dapple the thicket where she hides. When in danger, the fawn must stay very still. Any movement might draw the eye of a hunter.

The fawn and the doe, and the prowling lion—each one of these animals is camouflaged.

Mule Deer Fawn

What color is camouflage? Camouflage can be a certain color, or pattern of colors, or a special shape that fools the eye. Animal camouflage is a kind of disguise. It makes an animal hard to see.

Camouflage helps an animal hide from enemies, and it can help a hunter sneak up on its prey.

Hiding and hunting are both made easier when an animal matches its surroundings. A green insect clinging to a green leaf is much harder to see than a red insect. If a green insect is shaped like a leaf, even the hungriest bird could miss it.

Some animals disguise themselves by dressing in plants, pebbles, even other living things. Many crabs are experts at decorating their bodies with seaweed, shells, rocks, sponges, or anemones.

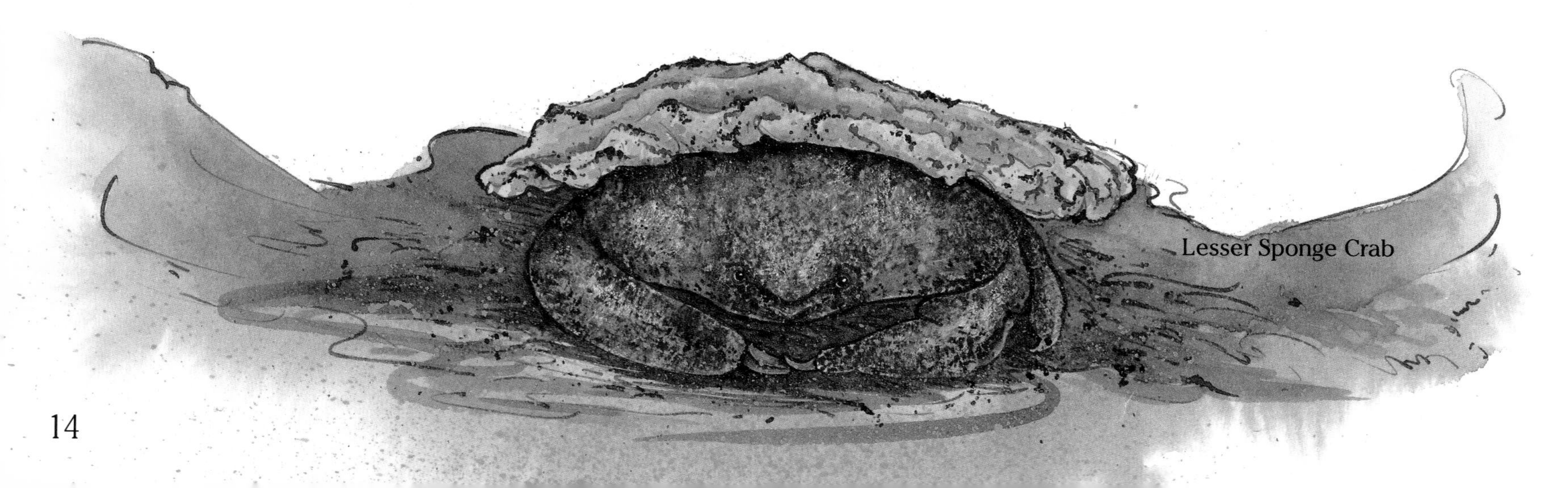

When animals are perfectly camouflaged, sticks seem to crawl, leaves can fly, and a stone may have eyes and a beak.

High up in the mountains above my house, this rock-colored ptarmigan warms her eggs. In summer, a ptarmigan has many dark feathers. By snowfall, her feathers will be pure white.

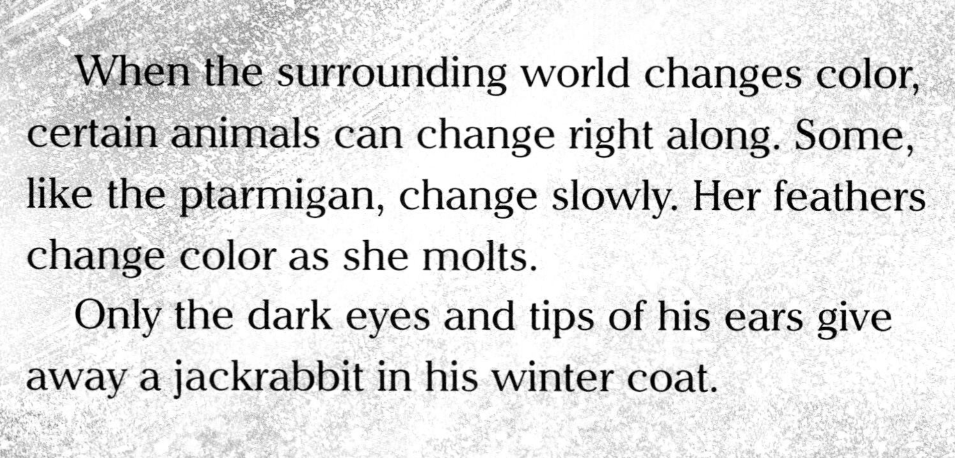

When the surrounding world changes color, certain animals can change right along. Some, like the ptarmigan, change slowly. Her feathers change color as she molts.

Only the dark eyes and tips of his ears give away a jackrabbit in his winter coat.

Young animals may change coloring as they grow—as they get big, or strong, or very swift.

Mule Deer

The fawn's spots will gradually fade, when she can run away from danger.

Though it may take nearly three minutes for an American anole to change coloring, certain chameleons, squids, and octopuses can turn different colors in seconds flat. They may change to match their surroundings, or because they feel excited or threatened.

Common Atlantic Octopus

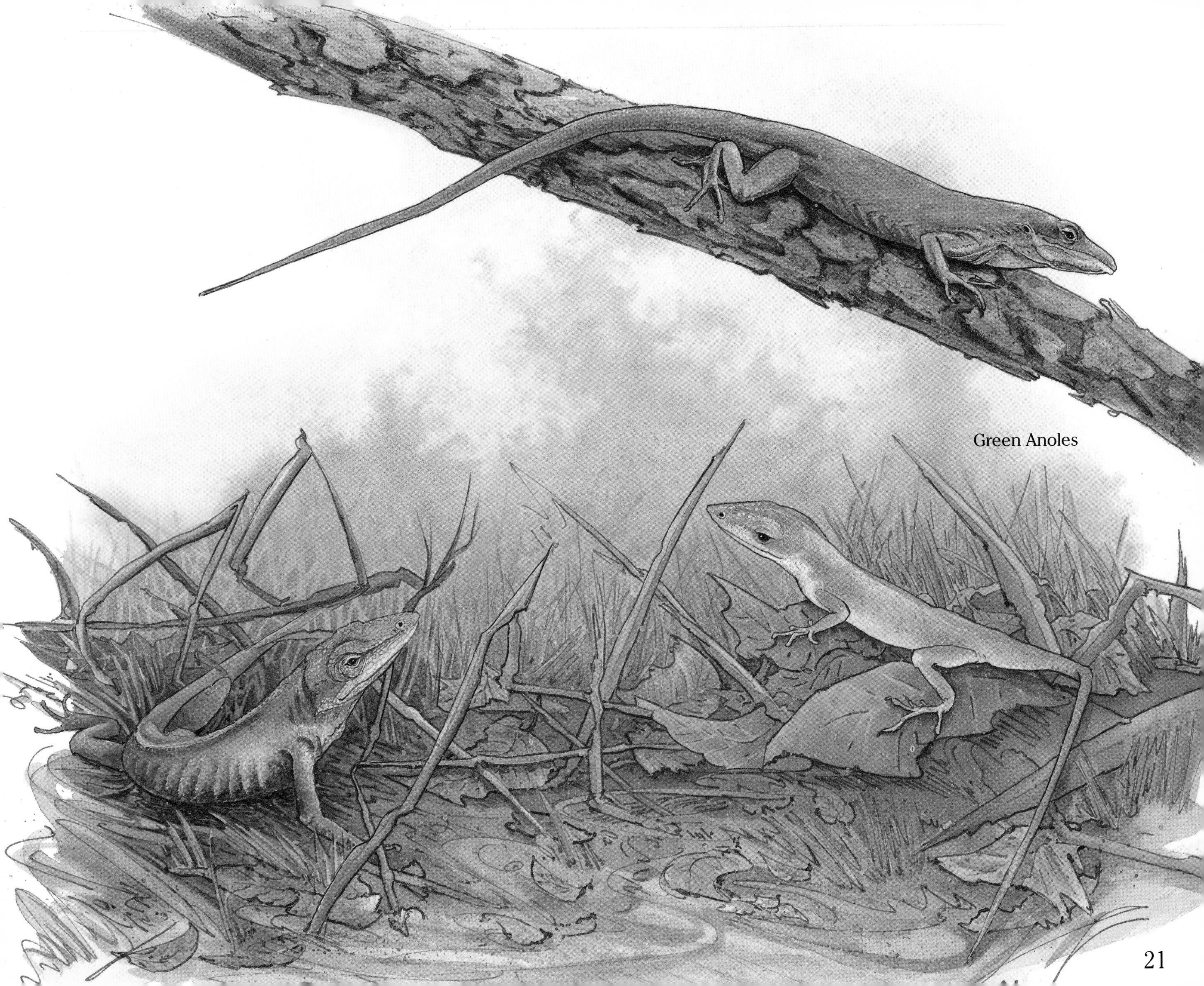

Green Anoles

American Robins

Many animals don't ever change. Some don't need any camouflage. These animals don't have to hide. Their colors may attract attention.

A robin's red breast helps him find a mate, and a skunk's bold stripes are a warning.

Animals sometimes use colors or bright patterns to say:

"I smell,"

"I sting,"

or "I'm poisonous."

Once sprayed by a skunk or stung by a bee, a hunter learns to avoid them next time.

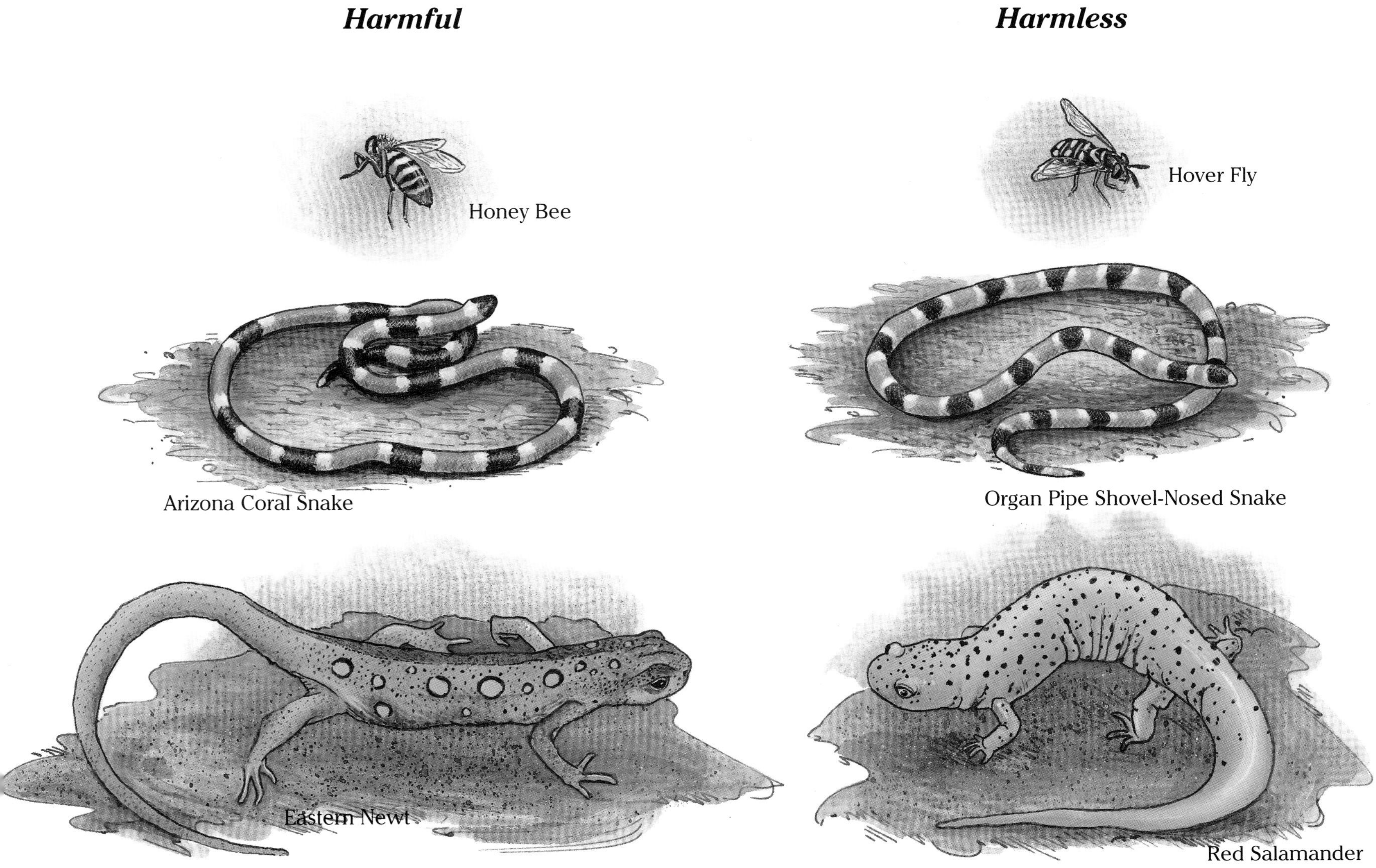

Harmful

Harmless

Honey Bee

Hover Fly

Arizona Coral Snake

Organ Pipe Shovel-Nosed Snake

Eastern Newt

Red Salamander

Harmless animals may copy warning signals to protect themselves from their enemies.

A flash of bright color may be enough to fool a predator, or to scare it away. Some animals have spots that look like eyes. Eyespots can frighten or confuse an enemy. A hunter may strike at "eyes" on a wing or tail, which gives the prey a small chance to escape.

Whether hunter or hunted, predator or prey,
colors and camouflage help an animal survive.

Queen Hornet Wasp

Wherever you live, animals live near you.

In backyards and in city parks, outside and inside your house, in plants, underwater, in air, animals are everywhere.

Do you see an animal?

An animal you can spot right away may be saying something important.

Black Widow Spider

Pine
Grosbeak
Spotted Skunk

Look closely, and more closely still. Look at colors, patterns, and shapes:

an old stump could be a bird,

a green branch may be a snake,

a twig might be a caterpillar,

a thorn might be an insect,

a rock may be a turtle,

and a dead leaf could be a toad.

Look closely when you go outside. Can you find the hidden animals?

Smooth Green Snake
Oak Inchworm Catterpillar
Flammulated
Owl
Snapping Turtle
Western Toad
Buffalo Treehopper

When animals are camouflaged, they can be very hard to see.

Otto, Carolyn.

What color is camouflage?

DATE